CORRECTIVE JAW SURGERY

A PATIENT'S COMPLETE GUIDE

ATTENTION

This book was originally designed to be used at The Center for Aesthetic & Corrective Jaw Surgery. Although most of the information is transferable, some of the descriptions, recommendations, and instructions may not apply to every patient.

Please consult with your doctor prior to following any recommendations or instructions in this book.

ISBN 978-1-300-81729-1

Corrective Jaw Surgery: A Patient's Complete Guide

Copyright © 2014 by John P. Amato, DDS, MD

All rights reserved. Except as permitted under the United States Copyright Act of 1976, no part of this publication may be reproduced, distributed, or transmitted in any form or by any means, including photocopying, recording, or other electronic or mechanical methods, without the prior written permission of the publisher. For permission requests, please email the publisher, addressed "Attention: Permissions Coordinator," at the email address below.

Email: info@TheCenterForJawSurgery.com

Ordering Information:

Special discounts are available on quantity purchases by doctors, associations, and others. For details, contact the publisher at the email address above.

CORRECTIVE JAW SURGERY

A PATIENT'S COMPLETE GUIDE

John P. Amato, DDS, MD

THIS BOOK IS DEDICATED TO

My family—for your undying love and support. Without all of you, nothing is possible.

All of the patients who have allowed me to help them throughout the years.

All of the orthodontists who make corrective jaw surgery possible.

Special Thanks To

John Christiana, my editor and graphic designer—for your incredible ability to bring to life everything I imagine.

www.johnchristiana.com

Dr. Cyrus Amato and Dr. John Amato performing model surgery in 1979

Dr. Cyrus Amato and Dr. John Amato in the operating room in 2009

"A good decision is based on knowledge."
– *Plato*

This book was created to give patients and their families the knowledge to make confident decisions throughout the jaw surgery journey.

It is designed as a reference guide and does not need to be read cover to cover. Read the chapters that apply to you first, but keep it nearby; new chapters will likely become relevant along the way.

Remember, you can do this! Be patient and stay focused. In the end, your life will be positively changed forever.

John P. Amato, DDS, MD

TABLE OF CONTENTS

INTRODUCTION

The Center for Aesthetic & Corrective Jaw Surgery

The Center for Aesthetic & Corrective Jaw Surgery, one of only a few such focused centers in the country, is dedicated exclusively to the treatment of facial bone imbalance affecting the jaws. The team of surgeons possesses a unique knowledge and experience of jaw surgery and has collectively treated thousands of patients throughout the country.

At The Center for Aesthetic & Corrective Jaw Surgery, patients experience a comprehensive approach to their facial bone imbalance. A patient's orthodontist is an integral part of the team and will typically work together with the surgeons to manage the treatment process for patients undergoing corrective jaw surgery. The surgeon and orthodontist work as a team throughout the process.

Once a patient is referred to the center, he or she will be evaluated by oral and maxillofacial surgeons, plastic and reconstructive surgeons, and craniofacial surgeons. In addition, the following services are provided in consultation as necessary:

- Speech therapy
- ENT surgery
- Nutrition counseling
- Physical therapy
- Patient advocacy for insurance coverage

The team of surgeons at The Center for Aesthetic & Corrective Jaw Surgery is dedicated to providing the most sophisticated level of care available for the treatment of facial bone imbalances. The team's focus on corrective jaw surgery over the years has led to the development of many exclusive techniques and devices.

Custom Devices Developed

- Custom Facial Analysis Software System—2012
- Custom Facial Growth Analysis Software System—2012
- Custom Skeletal Model Mounting Device—2005—U.S. patent
- Custom Facial Jaw Surgery Devices—2001—U.S. patent
- Custom Midface Distraction Device—2001
- Custom Jaw Widening Device—1999

At The Center for Aesthetic & Corrective Jaw Surgery, patients and their family members are considered an integral part of the team. Individuals are always encouraged to contact the team with any questions they may have.

For more information, please visit our website:
www.thecenterforjawsurgery.com